MW00509326

MORE ABOUT ANIMALS

BY MARGERY BIANCO

THE LITTLE WOODEN DOLL
Illustrated by Pamela Bianco

THE HOUSE THAT GREW SMALLER
Illustrated by Rachel Field

ALL ABOUT PETS
*Illustrated by G. Gilkison and
with photographs*

MORE ABOUT ANIMALS

WITH OTHER PUBLISHERS

THE VELVETEEN RABBIT
THE SKIN HORSE
POOR CECCO
THE HURDY-GURDY MAN

MORE ABOUT ANIMALS

BY

MARGERY BIANCO

Illustrated by
HELEN TORREY

NEW YORK
THE MACMILLAN COMPANY
1946

To

L. S. B.

CONTENTS

INTRODUCTION: HOW AND WHY   xi

RAINING RABBITS   1

BUSTER   8

THE CAT WHO WATCHED FOR THE MAILMAN   19

THE TRAVELING HORSE   26

THE GENTLEMAN IN BROWN   32

THE DOGS' FESTA   41

A QUEER NIGHT   47

ZINNIA AND HER BABIES   52

VISITORS   60

SPOT IN THE CONGO   69

JIM CROW   80

A BLACK-AND-WHITE BURGLAR   89

SPIKY AND CO.   97

COUNTRY NEIGHBORS   106

ILLUSTRATIONS

"PEOPLE," THE AUTHOR'S CAT   *Frontispiece*

                                             PAGE
RABBITS   I
BUSTER   II
BUDDIE AND THE MAIL   19
BOB, THE TRAVELER   27
WOODCHUCK   32
TROTTIE AT THE WOODCHUCK'S HOLE   37
ZORE AND BARBUN   41
OWLS   47
ZINNIA   52
LUNAR MOTH   63
SPOT   71
JIM CROW   83
SKUNK   91
HEDGEHOG   97
CHIPMUNK   106

HOW AND WHY

Some time ago I wrote a book about pets. I wanted very much to make it a useful book, more than anything else, but all the while I was writing it different recollections and stories kept coming into my mind, things that I wanted to tell about, and some of them got squeezed in and some didn't. For when one starts remembering, especially about animals, there is no end to it, and it is very easy just to ramble along. Still a good many people seemed to like the story parts of the book as well as the rest, and wanted to hear more of them. So it seemed a good idea to make a second book out of all those things that wouldn't go into the first one.

And here it is.

These are all true stories about certain animals I have known or met at various times. Some I know very well indeed; others have been companions during a summer or guests for a brief time, but whether our friendship has been long or short it has left in each case a very vivid memory.

Perhaps in a sense these are more recollections than

stories. Some of them are just funny incidents that happened, and that may amuse you just as they amused me and others at the time. For it is the same with animals as it is with people that you meet; what you remember best about them, very often, are just the little things, how they looked and the way they acted, and this does not always make a story in the real story-book sense. But it does make the kind of picture that comes instantly to your mind when someone says, "Do you remember Spot?" or, "Will you ever forget Buster!"

So what I have tried to do is to set down for you some of these things I remember, and to tell them just as they happened, or as they were told to me.

For though there may be all sorts of reasons and excuses for writing books, there can be only one excuse for writing this kind of book, and it can be put into just three words: I like animals!

—M. B.

RAINING RABBITS

One hears of it raining cats and dogs, but did you ever hear of it raining rabbits?

One spring morning—oddly enough it was just a day or so before Easter—I happened to look out from the front window, and I noticed that a little dog across the way was behaving very strangely. He was rushing about in circles and barking at something in the long grass by the fence, and whatever it was that he had found, it seemed to excite him very much. My first thought was that he had found a kitten, and that the kitten was probably not enjoying the fun nearly so much as the little dog. So I dropped my work and ran over to find out. But it wasn't a kitten at all; it was a fluffy half-grown white rabbit, all crouched up in a ball under a clump of day lilies, looking very frightened and bewildered.

I chased the little dog off, much to his disappointment, and picked the rabbit up. He was evidently a pet rabbit, for he seemed quite tame, and snuggled down in my arms as I carried him across the road. There he settled down contentedly with a carrot on the study floor while I set out to make inquiries.

No one, however, seemed to know anything about him, or where he came from. No one had seen or heard of a white rabbit anywhere in the neighborhood. So far as we could make out, he was just an Easter rabbit, and he had come from wherever it is that Easter rabbits do come from.

Now rabbits, even Easter rabbits, don't usually fall from the skies, but it certainly seemed that day as though they did, for when I went a little later to call on a neighbor who lived a few steps down the road I discovered her, much to my surprise, dodging about her vegetable garden on her hands and knees. It seemed a queer sort of performance, especially for her, and it reminded me somehow of the way the little white terrier had acted that same morning across the way, except of course that my neighbor, being a dignified kind of person, didn't bark. And when I said to her: "What do you think! I

2

just found the loveliest little rabbit, out there in the road!"
she just sat back on her heels and stared at me rather
bitterly.

"Rabbits!" she said. "It's *raining* rabbits! There are
two this minute right here in the garden, and it looks
as if they'd have every blade eaten before I can chase
them out!"

And with that she began dodging about again on all
fours, flapping her apron and crying "Shoo . . . shooo!"
while the two rabbits, a fat gray one and a spotted one,
lolloped about in and out the vegetable rows in that
provoking way that rabbits will, twitching their ears
and kicking their heels up, and pausing every moment
to snatch a lettuce leaf or a young beet top. Evidently
they were out on a picnic and thoroughly determined
to have a good time.

If you have ever tried to catch a tame rabbit who doesn't
want to be caught you will know just what we went
through that warm spring afternoon, chasing up and
down the garden with the rabbits always just one hop
ahead, moving as lazily as they could but always manag-
ing to escape just as one was about to lay hand on them.
While we were chasing those two, another rabbit sprang
up mysteriously from under our very feet, a black and
white one this time, and went scudding off among the

rhubarb plants. And while we were still staring at him, lo and behold another!

At this we really did begin to feel as though the garden—and in fact the entire neighborhood—had gone suddenly bewitched, beginning with my unexplained white rabbit in the road that morning. It was as if the whole landscape had begun to ooze rabbits, and one might expect them to pop out now in dozens, rabbits everywhere one looked.

My neighbor said: "Well, I give up. This is Easter, and no mistake!"

As a matter of fact there were only five, all told. But five were enough. They seemed more like twenty as they scampered here and there. By dint of much strategy and dodging, careful creeping up behind and sudden rushes at the last moment, in which we more than once bumped our heads together over the cabbages, we managed to capture four of them, and bore them off to the woodshed where we shut them safely in. The fifth, the black and white one, who seemed possessed of more devilishness than all the rest put together, was not to be caught by any mere human devices, so we chased him off into the far currant patch and left him to his own company.

The mystery was explained next morning. The rab-

4

bits belonged to a carpenter and odd-job man who lived alone in a little shack in the woods. When he went off to work that day he had left his barn door unlatched, and the rabbits had very sensibly taken advantage of his oversight to make their way, in a body, to the first thrifty-looking garden patch they could find.

Even this, however, did not explain my white rabbit. The white rabbit did not belong with the others. The carpenter had only brown and spotted ones. But he told me that a tame rabbit, kept alone, if it escapes from its hutch will sometimes travel for long distances until it comes to a place where there are other domestic rabbits, and that it seems to know by instinct in just which direction to go. It was possible, he said, that our little rabbit had come from several miles away.

Whether or no, nobody ever turned up to claim him, though we asked all the school children and also put a notice about him in the village postoffice. So he continued to make his home with us.

From being a small and dainty rabbit, a sort of *Alice in Wonderland* creature whom one might quite expect to see with a fan and kid gloves, he grew, in an amazingly short time, to be one of the biggest and lankiest rabbits I have ever seen. Somewhere in his family there must have been a strain of Belgian hare. By fall it seemed as

5

if he couldn't possibly grow any more. When he stretched out before the open fire—which he was very fond of doing of an evening, playing about the room till he was tired and then flopping suddenly full-length on the hearthrug—he looked exactly like a white woolly lamb.

As befitted an Easter rabbit, he was full of tricks and pranks, one being to snatch any small article of clothing, like stockings or underwear hung to air near the fire, and drag it down on the floor. He would even pull things out of bureau drawers that were left ajar. He stole slippers and hid them, but I never recall his doing any actual mischief, as rabbits too often will, beyond stealing his own dinner—and sometimes part of ours—from the vegetable bin in the shed. Any package that arrived he was immensely curious about, and gave one no peace till it was opened. Usually the contents did not interest him, but he had to know what was inside.

Housework he loved. He would follow a broom into every corner of the room and there seemed to be something about the rhythm of sweeping that excited him, so that he would hop and twirl and perform the most elaborate pirouettes, generally ending with a wild scamper round and round the floor.

He became so much a member of the family that we all felt very sad when the time came to move to the city, and our Easter rabbit went to live with two little girls in a cottage near the beach.

Another rabbit who appeared mysteriously in a garden, this time in a London suburb, caused quite a little excitement. He was a very pretty black and white rabbit, and was discovered devouring someone's most treasured violas. There was a great hue and cry about it, and after the rabbit had been finally cornered we offered to take him in and give him tea until his owner was found. We had a nice party, with sponge cake and radishes (the violas didn't seem to have spoiled his appetite at all) and the rabbit had the seat of honor on a sofa cushion in the middle of the floor. He behaved most decorously, made friends with the guinea pigs and drank milk from a doll's soup plate, and acted generally like a person of importance and real social training.

In the middle of the party came a frantic telephone call; were we the people who had found a black and white rabbit? Fifteen minutes later a car drew up outside the door and the chauffeur stepped out with a wicker basket. Our chance guest was the stage rabbit who appeared nightly in the London performance of *Peter Pan!*

BUSTER

Sometimes people adopt dogs, and sometimes dogs adopt people. One of the most determined cases of adoption by a dog that I ever knew was that of Buster.

I didn't blame Buster at all; in fact I had every sympathy with him. His own family were very nice people and quite kind to him, but they were dull. They were the sort of people who thought that if you fed a dog regularly and gave him a roof over his head that was all that he could possibly need.

We were living in an old roomy house on the outskirts of a village, and Buster's people lived in one of a little group of new bungalows that had sprung up just across the way. The family consisted of two grown sons who were away all day at business, and a stout middle-aged mother who divided her time between housework and listening to the radio, preferably nice cheerful jazz. Buster himself (we knew his name from hearing it called so often) was a lanky half-grown brown and white hound puppy, with floppy ears and enormous paws, built for romping. I think that the continual atmosphere of

8

scrubbing and polishing around the bungalow preyed on his mind, and I suspect that he did not at all care for radio. A good many hounds dislike music, especially jazz. He was a sociable dog, and what he wanted was someone to pass the time of day with. At any rate he decided that our family suited him a great deal better than his own.

He began by appearing casually at odd moments in the dooryard, just loitering about, in an expectant sort of way. It was useless to ignore him because sooner or later you were bound to catch his eye, and once you caught his eye the mischief was done. He would then begin to wag his tail and wiggle all over and act exactly as if you had called him in on purpose and were just as delighted about it as he was. Nothing discouraged him, and it was impossible to offend him.

For one thing his own family never went for walks, and we did. Buster loved walks. He liked long half-day tramps through the pine woods, and he never missed a chance to go along. If by any accident he did not see us set out he was sure to follow our trail and catch up, panting and triumphant, before we had gone a mile from home. If no one was going out walking, or down to the river, or doing any of those things he particularly enjoyed, then he would just hang wistfully about the yard.

9

And every evening, last thing, when I went to close the front door, I would hear a gentle tail-thumping from the darkest corner of the porch. It was Buster, settled down to guard our house for the night.

One always has rather a guilty feeling at the idea of having lured somebody else's dog away from them, though to be sure Buster never needed any luring—quite the contrary. So to salve our consciences we sent him home whenever we could, never an easy job, for he was singularly dense about commands of any sort; and we made it a strict rule never to feed him, in order that he should associate mealtimes at least with his own household. But he was so engaging, and so pathetically anxious for our company, that we were apt to be rather weak-minded about the whole business.

All of us, that is, except Grannie. Grannie was the only one who succeeded in being at all stern with Buster. She was very fond of dogs, but she liked them in their proper place, and she felt that a dog's proper place was guarding his own home. She believed, moreover, that if one really made it sufficiently plain to Buster that he belonged to the bungalow lady and not to us, he would end by understanding.

She would point a finger at him and say: "Now, Buster, you go *home!*"

And Buster, who was always anxious to please everyone, would sometimes actually go, though only to reappear again the moment her back was turned.

It was after one of these lectures, so often repeated,
that Grannie said: "I really do think I've taught that dog
at last where his proper home is, and now I hope he'll
stay there!"

The rest of us said nothing. We knew Buster.

Early next morning my daughter called me to the
door. She was laughing. "Look, there's Buster back
again! And what's more, he's bringing his bed with
him!"

There was Buster coming up the front path, looking
more than usually pleased with himself and dragging
with him, to our consternation, one of the bungalow
lady's best parlor rugs! It was a beautiful rug with
fringed ends and big pink and yellow roses on it, though
not at all improved by being dragged across a sandy
road and over the wet grass edges. He brought it straight
up on the porch, laid it down and sat on it, wagging
his tail and looking up at us very proudly, as if to say:
"There! Now aren't you pleased? Just feast your eyes
upon *that!*"

Whether he really meant it as a hint that he was moving over, bag and baggage, so that this tiresome question

13

of residence should be settled once for all, or whether he had caught the rug up in a sudden burst of generosity, wishing to make us a present of it, I don't know. Personally I think he intended it as a peace offering to Grannie, being in his opinion just the sort of gift calculated to soften and delight an old lady's heart.

We glanced guiltily over the way. Fortunately Buster's lady was invisible. Hurriedly we rolled the rug up, tiptoed across the road and dropped it in her front garden.

Buster looked puzzled and disappointed. He had brought us a beautiful present, why should we refuse it? Human beings, he sighed, were very strange. You did your best to please them, and there—they would have none of it!

For several days we went in terror lest Buster should bring us more gifts; a set of fish knives perhaps, or a bedspread. But evidently he had taken the rug incident very much to heart. He made no further advances of the kind.

Fall turned to winter, but not even the snow could keep Buster from our threshold. He still came for walks, accompanied the young people skating and sledding. And every evening, just as we settled down round the fire after supper, there would be a gentle scratching at the door. Buster had come for his usual visit. We let him stay until bedtime and then firmly put him out, the theory being that he then went home to sleep in his kennel, or whatever other accommodation his owners provided for him.

It was only on condition of his being sent home strictly at ten that Grannie would at all consent to his being allowed indoors. And he was so strangely good about going that Grannie remarked more than once: "You see, that dog can be perfectly obedient if only you are firm with him. He is trained now so that he will go home by himself with no trouble!"

It was just as well that she never noticed the wink Buster gave me as he went out of the front door and heard the key turned behind him.

For Buster and I had our own conspiracy, due to the coldness of those winter nights. He would patter noisily off the porch, stretch himself, listen a moment, and then creep silently back to the shelter of the big wistaria vine. There he would wait until he heard the house bolted for the night and saw, a little later, the light turned out in my room. I slept on the ground floor, and the windows were a few feet from the ground, just high enough for Buster to reach the sill with his front paws. When I pushed up the window, very quietly (Grannie slept in the room just above) there he was waiting. I would reach out and get a firm hold of his collar, there was a yank and a heave, and an armful of cold and snowy dog would come hurtling through the window. And if I wasn't very quick to get the window closed down again and make a dash for the bed Buster would be there before me, diving in snowy paws and all to bury himself as deep down under the blankets as he could get.

Of course it was all very wrong, but thin-coated dogs can't sleep out on a porch in winter, and nothing—nothing would induce Buster to sleep at his own house!

One thing puzzled me; that he was always so ready to go home first thing in the morning, almost ungratefully so, considering. If I was a few moments late in letting him out he seemed in a desperate hurry, and he always dashed straight across the road, looking neither to right nor left.

It seemed a good idea to call on Buster's lady, and find out tactfully where the dog really was supposed to sleep, so one morning I went over. I told her how much we all liked Buster and what a nice dog he was, and she said what a nice quiet dog he was, too, no trouble at all. And then we chatted a little and bye and bye I asked her, casually, where Buster slept at night. It seemed rather cold, I suggested, for a dog like that to stay outdoors. Maybe he had a kennel?

Indeed he had, Buster's lady assured me, and a nice kennel too, only he didn't seem to use it much. I could have told her that myself.

"But ever since the real cold weather set in," she continued, "he's taken to sleeping under the house here. We put some old sacks there for him to lie on and he seems perfectly comfortable. There's not a sound from him all night. Every morning when I open the kitchen door to call him in for breakfast he comes out from under the back porch, stretching and yawning like he'd had a real good night, and he's always just as warm and dry as can be!"

Not for worlds would I have given Buster's secret away. But I had found out his trick. The bungalow had no cellar, but was built on a brick foundation, open in places where a dog could creep through, and it was under here that Buster was supposed to sleep. Next morning, when I let him out before the rest of the household was astir, I watched carefully. Smoke was rising from the bungalow chimney; undoubtedly there were bacon and griddle cakes that moment on the stove. Buster made a straight leap across the road and I saw him squeeze himself under the front porch, only to crawl out next moment,

I felt sure, at the back, yawning and stretching as the bungalow lady had described, all ready for the breakfast that was awaiting him.

There was no need to worry about Buster! A dog as clever as that could take care of himself anywhere.

THE CAT WHO WATCHED FOR
THE MAILMAN

Buddie was a funny-looking, rather than a pretty cat. He was white, with black spots and markings, and the markings were so placed that they made him look rather like a clown. He began life as an alley cat, and like many alley cats he was intelligent in peculiar ways.

Even as a kitten he showed good judgment in choosing the people he wanted to live with. He simply went one morning and sat on the landing outside their front door.

It was quite early; the occupants of the apartment, two ladies, were going out for the day and they were in rather a hurry. They saw the little kitten sitting there, looking hungry and bedraggled, but they had no time to look after him then. So they said: "We can't do anything for you just now, kitty, but if you'd like to stay around and wait for us, we'll see what can be done to-night."

Buddie stayed. When anyone shooed him away from the landing, he went right back.

In the evening, when the family came home, the janitor told them: "That little kitten has been waiting around all day for you, and I can't chase him away. If you don't want him I'll put him outside in the street."

They said: "That's all right. We told him to wait." They unlocked the front door, and Buddie marched in. He ate his supper and drank his milk and then tucked his front paws under him and settled down to stay. He had no doubt at all but that his new friends would be quite willing to keep him, and they did.

Later on, when the family moved to Washington, Buddie went with them. There they lived in a house with a garden, instead of an apartment, and Buddie was free to come and go as he chose. He was a big full-grown cat by now, and he liked to go out on hunting and exploring excursions in the neighborhood. There was a big park not far from the house, parts of which were quite wild, and it was there that Buddie liked to wander. The family worried about these excursions, for they knew that in fall and winter the boys often set snares and traps in the wilder parts of the park, where the warden was not likely to catch them. Buddie was a clever cat, but even to clever cats bad things will sometimes happen.

Buddie often stayed away for a day or a night, but he always came back sooner or later. One winter however, when it was cold and snowy, he was absent for five days, and his mistress was very much afraid that he was lost for good. But on the sixth day, quite late, she heard mewing, the particular clear mew of a cat who is calling for help. She ran to the door, and there was Buddie, dragging himself very slowly up the garden path and mewing as he came. There was a sound of something clinking, too, on the frozen path. When he came into the light she saw that he had a steel trap, with a bit of broken chain trailing from it, fast to one of his front paws.

He was nearly frozen, very thin and exhausted from pain, but he stopped mewing when she picked him up, and sat perfectly still on her lap while they pried the trap open to release him, and bathed and bound his broken paw. He must have come a long distance dragging that heavy trap with him, over rough ground and across fences and up the hill finally to the house, a long and painful journey. His foot was in such bad shape that they were afraid at first he would have to lose it, but at last it healed, and he was able to limp about on it once more. After this accident however he stayed nearer home, and gave up his long hunting trips.

There was a friend staying in the house at the time, who helped to nurse Buddie and dress his injured paw every day. Buddie became very attached to him—even more than to his own family—and would spend all the time he could in his company. This friend was a writer and, while he worked, Buddie with his bandaged paw used to lie on his desk watching him.

Buddie's foot healed and the friend, whom we will call Charles, went back to his own home. But he often came to visit at the house, and Buddie always seemed to know when he was expected. Perhaps he recognized his step a long way down the street, as cats will—they have very acute hearing—for always, several moments before the door bell rang, Buddie would be sitting there on the hall mat ready to welcome him.

Presently these visits ceased, for Buddie's friend went abroad. He was away for more than a year, and during that time he wrote to Buddie's family, sometimes frequently, sometimes at long intervals, so that there was no regularity about the arrival of his letters. But somehow Buddie always knew. Usually he took no interest at all in the mail. But if, on any morning, there was a letter from Charles among the envelopes lying on the hall mat Buddie would be found there, keeping guard over it. He would wait until the letters were gathered up, and then he would follow to his mistress' room, a thing he never did on other mornings, and sit close beside her while she read her mail. He seemed to watch her face, to see whether the news was good or bad, and when he had been reassured about this his mistress would lay letter and envelope down, and Buddie would curl right down on it, his paws tucked under him, and lie there contentedly purring.

How Buddie ever knew that certain letters were from his friend is a mystery. He could scarcely have recognized the writing, but it is possible that his sense of perception was so acute that he could distinguish an envelope which had been handled by someone he knew and loved, even though it had traveled a long distance and been mixed up with so many other envelopes in the mail bag. That at least is the only explanation that seems possible, and it is quite certain that he never once made a mistake, and that he never missed any letter that arrived. Whenever the family saw Buddie sitting in the hall just after the mailman arrived they would say: "Why, there must be a letter from Charles!" And there always was.

The return of this friend whom he loved so much meant great rejoicing for Buddie. Now everything would be wonderful again. For a time it was. Then Buddie's friend fell ill. He had to go into hospital for an operation.

Buddie knew, through that mysterious extra sense of his, that something was wrong. He had no means of knowing what it was. But he was aware that this absence of Charles was not like his other absences. He began to worry. He went about the house restless and mewing; studied the faces of the family as they came and went, and even tried to follow them when they left the house. He did everything but actually ask in words what the trouble was.

Finally he worked himself into such a state that the family were quite concerned about him. They tried to explain, but verbal explanations of course were of no use. Buddie had to see for himself, and be satisfied.

Dogs and cats are not allowed in hospitals as a rule, but the doctor happened to be a good friend and so a special exception was made in Buddie's case. He was taken to the hospital, carried up in the elevator and into his friend's room.

The invalid stroked him, and said: "It's all right, Buddie. I'm going to get well again. Everything's quite all right!"

Buddie studied his face very earnestly. Then he examined everything about the room. What he saw seemed to satisfy him. He allowed himself to be carried home again, with no resistance, and once there he settled down at ease. There was no more restlessness, no more prowling and mewing. Buddie had seen for himself that his friend was alive, that he was being taken care of properly, and now his mind was at rest and he was content to remain tranquilly at home and wait for his return.

THE TRAVELING HORSE

Some friends who lived in the country many years ago once wanted to buy a horse. They already had one horse who did the hauling and farm work, a nice slow-footed old creature, who was very dependable but not much fun to drive around with. So they thought they would look about for another, a younger and smarter horse to use on the road.

Among the many horses brought for their approval was one to which they took a particular fancy, a very pretty little sorrel with a light mane and tail, called Bob. He seemed very willing and gentle, and they decided to buy him.

For some time everything went well. They took long drives all about the countryside, which the little horse seemed to enjoy just as much as they did. He was a perfect little driving horse, and he looked so sleek and smart between the shafts that they felt very proud of him.

But after two or three months, to their surprise, Bob began to be troublesome. He grew lazy and careless. One fine day, when they went to harness him, he be-

haved just as badly as he could. First he stood stock-still, planted his four feet firmly and refused to budge one inch. When they tried to lead him, then he began to back and rear and toss his head, and act altogether as though he had never been hitched to a wagon before and was determined to get rid of it at all costs.

His owners could not imagine what was wrong. They examined the harness, and the wagon too, but everything seemed all right and just as usual. They tried coaxing, and they tried firmness, but nothing would induce the little horse to behave. So in the end they had to put him back in the stable and give up their drive.

The next time they tried to hitch him up, the same thing happened. He just refused to be driven.

Now a horse that you cannot drive is as good as no horse at all, and his owners began to be really worried. And while they were still wondering where the trouble lay, and what could be done about it, a man who lived in that neighborhood, and who bought and sold horses, happened to call at the house.

They told him what trouble they were having with the little sorrel, and the man began to laugh.

"Let's have a look at your horse," he said, "because I have an idea I know him."

They took him to the stable. As soon as he stepped

inside the little horse turned his head to stare at him, and it was almost as if he winked!

"Yes," said the visitor, "that's the horse, right enough. He knows me and I know him! He's been through my hands more than once, and now it looks as if I'll have to sell him again. How long have you had him, three months?"

"Just about that."

"Well, you'd better let me make a trade with you, for I can tell you one thing; you won't be able to drive that horse again, take my word for it."

And then he told them the little sorrel's history.

It seemed that the horse had been owned, at one time or another, by nearly everyone in the district. The mailman had had him, the baker had had him, several farmers had had him. But no one had been able to keep him for more than three or four months, and that for a peculiar reason. The horse was in perfect health, he wasn't vicious and he couldn't be said to have a single fault—except one.

He just liked to travel around.

When he first went to any new place he behaved perfectly; a better horse couldn't be found. But sooner or later he would begin to get restless; he was tired of that place and he wanted a change. And his way of getting it

was always the same. He deliberately set himself to behave so badly that his owner would have to get rid of
him, and send him somewhere else. And in this way he
had managed to change hands so often that his character
was known for miles around, except of course by our
friends, who happened to be newcomers.

Of course, hearing this, it was useless for them to try
and keep the little horse any longer, much as they liked
him.

So away he went again, trotting off quite contentedly
at the tail of the horse dealer's wagon, tossing his mane
and seeming very pleased that he had succeeded in getting his own way once more!

Somewhere or other he is probably still going his
rounds, carrying out his own idea of how an enterprising
horse should live without boredom.

THE GENTLEMAN IN BROWN

There are some animals that no one seems to like. The woodchuck is one of them. There are many who admire the skunk, in spite of his drawbacks; charming poems have been written about the mole whom gardeners so detest. But I have seldom found anyone to say a good word for woodchucks.

It is true that the woodchuck is clumsy, impudent and homely. His walk is a waddle; he is snub nosed, with bristles on his face, and in figure a portly woodchuck—and most woodchucks are portly—resembles nothing so much as a shabby brown plush sofa cushion with a leg at each corner.

Perhaps if he were more graceful, with a slender nose and a feathery tail, people would be less inclined to give him such a bad character!

Only once a year, on Candlemas Day, does the wood-chuck get any sort of recognition, and then he is usually referred to as the "ground hog." On the second of February, according to tradition, he comes out blinking from his winter sleep to look for his shadow, and if he sees it, then winter will last another six weeks. So once a year, at least, he has his picture in the papers, and that should make up a little for the mean things that are said about him during the rest of the twelve months, especially in early summer, when everyone's thoughts are on the vegetable garden.

Year after year I had heard so much about wood-chucks, how greedy they were and how much damage they did among the peas and beans and young cauliflow-ers, that it almost discouraged me from ever attempt-ing to have a garden at all. Everyone had her tale of woe; everyone had new plans each spring for outwitting the woodchucks, and none of them, it seemed, was suc-cessful!

It was all the worse because, this particular sum-mer, we had rented an old farmhouse that had stood empty for several years. And during that time, of course, the wild creatures had been having it all their own way.

Neighbors said cheerfully: "Why, you'll never keep a

garden there! The place is just *running* with woodchucks! They'll eat you out of house and home!"

And woodchucks there were, in plenty, though it was
some time before I actually saw one. Meantime I dug
and raked and planted my beans and carrots recklessly.
If woodchucks were going to come, let them come. There
should be plenty for me and them too!

For I have long held a private theory—which friends
make fun of—that as long as you let other creatures
alone, they will let you alone. Complain, and they will
give you something to complain about. Of course I
have never had a chance to try this out with anything
of a really fierce nature, but it certainly seems to work
with the smaller beings. How otherwise can you explain
that winter when mice invaded our home, and the only
person who shooed at them and said unkind things about
them, and baited quite useless traps to set in every corner of her room, was the only one whom they really
annoyed, and on whom they took revenge by biting
large holes in her most treasured sweater? Or the summer when yellow jackets built their nest under the porch
step, and the one person stung by them was a visitor who
hated wasps and insisted on flapping newspapers at them
every time they appeared.

But to return to the woodchucks.

34

One side of the garden was bounded by an old stone wall, and near this stood a small unused building that had once been a blacksmith shop. Needing one morning to find a bit of old iron for some purpose or other, I thought of looking in this little shed. No sooner had I opened the door than I was startled by a sudden shrill whistle—much like the sound boys produce by blowing on two fingers, only louder and shriller. Somewhere under the broken flooring one of the many Mrs. Woodchucks had made her home, and this police-like whistle was just her warning to keep away.

Anyone who for the first time hears an angry woodchuck whistle, close at hand, will receive just such a shock as I did, and that is exactly what the woodchuck intends. Coming so unexpectedly on the stillness, it is really quite a startling sound, and it carries a long distance. Woodchucks use it not only to scare off intruders, but also as a danger signal to one another. If ever you come upon a group of them on some old pasture slope of a sunny morning, sitting up outside their burrows as they love to do, just try whistling sharply on your fingers, and you will see them all fly to cover.

I soon learned to know that sound very well, for I heard it quite often, nearly every time I approached the stone wall, and although it didn't startle me again so much as the first time, still I never quite got used to it.

Inquisitive eyes must have watched my planting and raking, though my brown neighbors were at first very shy about showing themselves. Early morning was the time they chose for their inspection. Looking from my bedroom window I would see a fat brown figure hoist himself carefully from some hole in the stone wall, waddle down the path and examine those tiny green seedlings with the utmost interest. Up and down the rows he would go, sniffing and twitching his whiskers, with very much the air of a stout elderly gentleman out early to see how his garden was getting on. Sometimes Mrs. Woodchuck would join him, and they would have a consultation, sitting up solemnly on their rear ends and glancing now and then towards the house.

Bye and bye I caught glimpses of the young ones too, stealing in and out the thicket of goldenglow, or trotting over to the apple tree by the kitchen door to see if there were an apple or two for the picking up.

But except for the fallen apples, which later in the summer they gathered greedily, not one thing in the garden did those woodchucks touch!

Either they had too much of their natural wild food, whatever it may be, or else they decided that I was such

36

an amateur gardener that it wasn't even sport to rob me.

Not so the cutworms and other insects. They did their best to ruin everything. But then, as I waged war on them, it was quite fair on both sides.

One other person besides myself took great interest

in the woodchucks, and that was Trotty, the little Scotch terrier who then shared our home.

Her first sight of them was almost disastrous, for she was staring out from an upper window early one morning, and in her eagerness to discover just what that fat brown creature was, wandering about the vegetable garden, she overbalanced and tumbled straight out on her little black nose, much to the woodchuck's amazement.

37

For though he probably kept an eye trained for possible dogs on the ground, he could scarcely have expected to see one fall from the skies like that. Luckily no harm was done, but Trotty was just as astonished as the woodchuck at her own mishap.

I can't say that they ever became good friends, but Trotty learned not to interfere with them within the garden limits, at least. She did a good deal of fussy patrolling, as little dogs will, and would often lie out there in the blazing sun, when she would have been far more comfortable in the shade, just to show those impudent creatures that she had as much right to the garden as they had. But it never came to an actual encounter.

Certain plans, however, must have been brewing all the while inside her little black head, only waiting a chance to be put into practice.

On the pasture slope leading down to the brook, some distance from the house, lived another colony of woodchucks, who had made their burrows in the middle of a briar patch. Trotty, very likely because I had discouraged her from worrying their cousins who lived in the garden, was very inquisitive about this particular woodchuck family. Whenever we neared the briar patch she always tried either to rush on ahead, or to lag behind us unnoticed, and one day she succeeded. I heard a scuffle and a frightened squeal, and realized to my dismay that Trotty had actually caught one of the babies, who had wandered too far from the burrow.

Now woodchucks will fight very savagely in defense of their young ones, and Trotty might have had a very unpleasant surprise in another moment, when Mrs. Woodchuck came to the rescue, but luckily as I came up the baby managed to wriggle from between her paws, and scamper home, more scared than hurt. I thought, however, that a little explanation would be good, so I haled the young lady to the mouth of the burrow, pointed to it, and scolded her well.

"Bad dog!" I said, and immediately a gruff voice from inside the burrow added: *"Bobble*-bobble-bobble!"

It was Mrs. Woodchuck, very indignant, and determined to have her own little say in the matter too.

I was rather surprised, and so was Trotty. She looked in a puzzled way at the burrow, and then at me.

We waited a moment. Silence from the woodchuck hole. Then I said firmly and distinctly: "Very naughty, unpleasant little dog!"

And again Mrs. Woodchuck added: "Bobble-bobble-bobble!" this time less angrily, but as if she were saying: "Just what I think! I told you so!"

Being scolded by two people at once was just too

39

much for one little dog. Trotty hung her ears, and it was my turn to call into the burrow: "Look here, you've said quite enough, now. Don't be so unpleasant about it!"

"*Gr-rr!*" returned Mrs. Woodchuck promptly.

We felt, like Alice in Wonderland, that this last piece of rudeness was more than we could bear. We turned our backs on the woodchuck hole. Little dogs may sometimes act hastily but, after all, manners are manners!

Still, what can one expect from a woodchuck!

THE DOGS' FESTA

Our first summer in Italy was spent in a house among the hills just outside Turin. It was an old-fashioned country house, and in a smaller cottage near by lived the *contadini,* a man and his wife who took care of the cows and the vineyard, and from whom we bought our milk and eggs and fruit. They had two farm dogs, who were named Zore and Barbun.

Zore was a huge yellow mongrel, just outgrowing the puppy stage. He was as big as a young calf, clumsy, stupid, but very friendly with everyone. He had no sense at all, but a great fund of cheerfulness and affection. Barbun, his companion, was quite old. He was a strange-

41

looking dog, thin and grizzled, with a tangled coat like an English sheep dog and a much bewhiskered face, which was how he came to get his name, for *barbun* in the Piedmontese dialect means "bearded."

Barbun was very wise, very wary, slow to make friends but singularly gentle in his ways. While Zore would prance around one wagging his tail and coaxing to be played with, Barbun usually sat at a distance, serious and watchful, his eyes blinking through his tousled hair. I have said that he was thin, but I think that he weighed, for his size, practically nothing. He gave the impression, when one touched him, that his bones must be hollow like a bird's, for he seemed to have no weight or substance at all; it was as if a strong wind might have blown him away.

These two dogs were close companions; one rarely saw them apart. When they were not racing through the meadows, or helping to bring the cows in, they haunted our back doorstep, chiefly for food.

There was a big old apricot tree just behind the house, the haunt of nightingales on summer evenings, and under this tree the two would take their post, waiting for scraps. They were the hungriest dogs I ever saw; one might almost think they *were* fed on dish water, as the cook always insisted! Nothing came amiss to them.

Stale bread vanished at a gulp; they would devour melon rinds, egg shells, even coffee grounds, though Zore looked fat enough and I think Barbun's thinness was natural, for he was less ravenous than his companion.

It happened after some weeks that my husband had to go to Naples on a business trip. Every city in Italy has its own specialties in food, and one of the specialties of Naples is a sort of open pie called *pizza*. It is rather a queer delicacy, being made of cheese and onions and tomatoes, baked together in an open crust, something like a pumpkin pie, but the Neapolitans are very fond of it, and all through the city you find shops that make nothing else. They are baked in all sizes, from very small ones up to great big ones that would easily serve a dozen people.

When my husband passed one of these shops and saw the splendid array in the window he thought he would buy a *pizza* just for fun and send it back for us to taste. So he chose the biggest and finest one he could see— *pizza* is a popular dish and not at all expensive for its size—and he had it packed up then and there and sent off by mail.

Up at the villa there was no parcel delivery; the mailman had enough to do climbing that steep hill each morning with his letter bag. So all we got was a notice

43

saying that there was a package down at the postoffice which we must claim personally, and by some mistake there was no mention of the package being perishable. It was a long hot walk down to the postoffice and back, and for several days no one happened to go down. Everyone kept saying, "We *must* fetch that parcel," and still no one went. Then one evening my brother-in-law remembered it, and he brought it home.

We opened the box, undid the ribbons and the frilly lace paper—and there lay our *pizza!*

It was in August, a very warm season, and in addition to the train journey that parcel had lain in the postoffice for nearly five days. None of us had ever seen a real Neapolitan *pizza* before, but somehow we felt instinctively that it shouldn't look just like *that*. It was enormous, nearly two feet across, baked originally to a crispy brown, with golden cheese and red tomatoes, but by this time it was all colors of the rainbow! A strange greenish hue had crept over the yellow, and there were little forests here and there of blue and purple mold.

We looked at one another. Our first thought was to bury it, and as rapidly as possible, only that it looked such an enormous thing to bury. And then my brother-in-law had an idea. He said: "Let's make a real festa for Zore and Barbun!"

I felt sure it would poison them both, but he reassured me. Nothing, he said, in the way of food could possibly poison a *contadino* dog!

So, very gingerly, we picked the *pizza* up and carried it out under the apricot tree, and with some misgiving I called the dogs.

They came at once, wagging their tails. We pointed to the pie. They, too, had never seen a *pizza* before, and at first they were puzzled. Instinct told them that this was food for *signori*, not for dogs. They hung their ears and looked actually embarrassed. It was intact; no knife had cut it. Nothing so savory and so beautiful could possibly be for them!

But we explained. It was theirs, a gift; we gave it to them freely.

Then Zore sniffed the cheese. His eyes began to bulge and his great jaws opened. He snatched at one edge of the pastry, and Barbun caught the other. In one second they had torn it apart.

We left them to their feast. But even across the garden we could still hear Zore's eager gulps. When we returned, a little later, the beautiful *pizza* had disappeared. Not a vestige remained.

Barbun lay stretched under the apricot tree, his eyes dreamy, his tail faintly wagging. Near him sat Zore,

45

visibly bigger to the eye, sighing blissfully and licking a last minute shred of pastry from his paws.

For months after, I am sure, those dogs went about the neighborhood boasting to all their friends of the strange foreign *signori* and the wonderful festa they had made for them!

A QUEER NIGHT

Have you ever noticed that there are days when everything seems to happen in a queer way? I don't mean days when things go wrong, for that usually depends upon oneself, but just peculiar days. Things disappear mysteriously; you lay something down and the next minute you can't find it, or it turns up in some totally unexpected place where you are quite sure you never put it. If you live in the country the milk will turn sour and the potatoes burn, and you are very likely to find your pantry suddenly invaded by hosts of large black ants where you never saw an ant before. And above all, the animals behave in the strangest ways. As old-fashioned people would say, everything is plain bewitched.

I remember particularly one such day, or rather evening, when we were living in the house where there were so many woodchucks. The woodchucks had nothing specially to do with it; they went about their business as usual. I think woodchucks are about the last animals to feel any sort of bewitchment; they are far too solemn and matter-of-fact.

It had been a sultry sort of afternoon, with a thunderstorm threatening but never coming to grips. The air was heavy and breathless, and yet tiny whirlwinds seemed to spring up from nowhere all of a sudden, swirl a few straws and leaves about and then subside just as mysteriously. Bats were flitting about the lawn. It seemed a regular witches' evening.

Pick, the yellow collie, and little Scottie were unwontedly restless. They kept hearing things that no one else heard, pricking their ears to listen and then rushing off, only to return next moment and begin the performance all over again; a most irritating way for dogs to behave when you are sitting on the porch of a lonely farmhouse on a particularly dark and spooky sort of evening. And People, our black and white cat, seated in the square of lamplight from the window, was just as bad. He kept washing behind his ears—always a sign of something afoot, as country people will tell you—but his mind wasn't on his washing at all. It was only a pretense, for

48

every little while he would pause with a paw in mid-air and listen too, and you could see that the fur on his back was all electric and ready to bristle.

Then the whinnying owls began, those funny little owls that are just like ghostly voices. There was usually one somewhere in the branches of the big catalpa tree across the lawn, but tonight there seemed dozens of them. They kept calling to and fro in their clear hollow voices, with that little rising inflection at the end: "Hoo . . . hoo-*oo?* Hoo-*oo?* Hoo!"

For a time we listened to their eerie chorus, and then my daughter said half-jokingly: "I don't like it out here any more. I'm going indoors!"

"Nonsense!" I returned, feeling just a little bit spooky myself. "We get owls like that every night."

"*Do* we?" she said.

We sat listening. All at once, People sprang suddenly up, took a flying leap across both our laps, every hair on end and his tail like a broomstick, and disappeared into the darkness. We heard sudden blood-curdling yowls (not from People, who is a silent fighter) and the thud of racing paws. The howls retreated, grew fainter and fainter in the distance. After an interval People reappeared, as abruptly as he had gone, sat down on the porch again and resumed his washing.

The owls, for some reason, had stopped their hooting. There was the kind of silence in which one could have heard the proverbial pin drop.

Presently I became aware that People was watching, with a very intent and interested expression, something which was apparently taking place behind me. As I was sitting flat on the porch floor, my back against the wall, this was puzzling. And just then I caught my daughter's eyes, too, fixed in a very peculiar way on the floor boards near my back. She said: "Don't move. There's an enormous spider crawling by just behind you."

Now to say that was like telling someone who hates snakes not to move because a blacksnake is just gliding over him. For the one—and I trust the only—living thing which has power to produce cold shivers up and down my back is a spider. I know that spiders are intelligent, interesting and even beautiful creatures and I never want to injure one, but still the fact remains I can't abide them. I cannot even look at a spider without feeling very queer inside me. I kept still, mainly because I had no means of knowing just exactly where that spider was, but it took a great effort. And when she exclaimed: "For goodness' sake, there's another!" I sprang to my feet and whirled round.

There was the first spider, just letting himself down

in a leisurely way over the edge of the porch step, and there behind him, just crossing the square of lamplight, was the second, crawling deliberately along in his tracks. And further back still, in the shadows, other vague forms seemed to move and stir.

Gingerly I reached for the flashlight. Spiders—huge black spiders—coming along in Indian file, one behind the other and about a foot apart, swerving neither to right nor left, a strange uncanny procession! Each turned the corner at the angle of the wall and each, when he came to the step, hoisted himself down and set off along the garden path. Seven we counted, and there seemed to be more coming.

But we didn't wait to see the end of this singular parade. We looked at one another, and without a word made for the doorway and the lighted room. We had had quite enough for one night.

But where those spiders had come from, where they were all going, remains a mystery to me to this day. Unless, of course, they were going to the Witch's Sabbath!

ZINNIA AND HER BABIES

Some time ago, in writing about cats, I spoke of Zinnia, the little black cat who has such a liking for dogs. I told how she will sit staring pensively at some dog which may happen to be in the room, and then all at once jump up, walk over to him and begin washing his face and smoothing his hair, just like a nursemaid, holding his head between her two paws if he tries to move. And what a terrible time she had trying to straighten out the big long-haired collie! Much as she likes dogs, their occasional untidyness gets very much upon Zinnia's mind, and sooner or later she decides that something must be done about it.

I think Zinnia made up her mind long ago that all this so-called "cat and dog" business was just nonsense. Dogs are just like cats, if you treat them the right way, and a little good sound cat training would do them all the good in the world. Zinnia's owners are always afraid that she will one day, in her fearlessness, happen upon the wrong dog, and then she may get unpleasantly surprised. But so far all the dogs she has met have either returned her friendliness, or been far too surprised by it to think of attacking her.

The only other creatures that Zinnia seems to like better than dogs are—not human beings, for she is quite indifferent to petting or attention; and not even other cats, whom she just tolerates, but kittens. Anybody's kittens, though naturally she prefers her own. She is a born mother, and the voice of a kitten mewing will always bring her on a trot to see what the trouble is. Quite likely she may decide that the kittens are fretful because they are not in what she considers a good and comfortable place for them. In that case she will probably pick them up one by one and carry them off somewhere else, without bothering to consult their own mother at all.

One summer I remember Zinnia was bringing up a family of her own, four babies of a few weeks old, and she had them in the barn. There was a good deal of painting and other work going on about the house at that time, and Zinnia's mistress thought that the barn was the safest place for the kittens to stay in, at least until they were a

53

little older, when there would be less danger of their tumbling into paint pots, or getting stepped upon by accident. So the barn door was kept closed, and Zinnia's own food and saucers of milk were carried out to her there twice a day.

Zinnia, however, had her own ideas. The barn was big and airy; the kittens had their own box lined with hay and plenty of space to crawl safely about in. But that didn't satisfy her. It was sunny June weather; kittens should be out in the garden, enjoying the fresh air.

And every morning there they were, all four of them, packed into an old wheelbarrow by the woodshed, where they whimpered and blinked at the sunlight, looking, poor mites, anything but comfortable on the bare earthy boards. And there Zinnia would leave them, while she went off on some hunting excursion of her own.

For a long while no one could discover how Zinnia managed to bring those kittens out. Every evening, and often during the day as well, they were carried back to their own bed and the barn door closed and bolted, with Zinnia and her babies inside. But it made no difference. Every morning, there they were back again in the wheelbarrow.

But at last we found out Zinnia's secret.

54

High up in the barn wall, near the rafters, was a window with one small pane missing. Because of the slope of the ground, the barn was so built as to be two stories high at the back instead of one, so that from the outside this window was even higher from the ground, some fourteen feet or more. One morning Zinnia's mistress, passing near the barn, heard a queer scraping sound, like the noise a squirrel makes scrambling down a tree trunk, and looked up to see Zinnia, who had just squeezed through the broken window with a kitten in her mouth and was letting herself down cautiously, backwards, inch by inch, clinging with her claws to the smooth side of the barn.

The kitten safely landed on the ground, and dragged over to the wheelbarrow, Zinnia turned right around and went back for the next, by the same route.

Seeing that she was so determined about it, it seemed useless to shut the kittens up any longer, so after that the barn door was left ajar and Zinnia was free to carry her babies in and out in a less dangerous way. It was just like obstinate little Zinnia, however, that having once got her own way in the matter she didn't seem to care any longer whether the kittens were indoors or out, but left them most of the time to crawl about by themselves on the barn floor, where they certainly seemed happier

55

than when they were being dragged to and fro so ruthlessly every day.

All but one, the smallest of the four. Nearly always, if there happens to be one kitten in a litter smaller and weaker than the rest, the mother cat will choose it as her special favorite, and so it was with Zinnia. This particular kitten, a tiny maltese with a white chin, Zinnia took extra pains with. It was slower in learning to crawl and eat, due, everyone thought, to having been carried about so persistently by its mother, and perhaps getting a few bumps here and there sliding down the barn wall. Every day, long after she had left the other kittens to their own devices, Zinnia would still drag this weakly one out into the sunshine, whether or no, and leave it there for several hours in the wheelbarrow, and either her devotion, or the warm sunshine, had its good effect, for the little malty grew up in time just as strong as her brothers and sisters, though always a bit smaller.

But the queerest thing that Zinnia ever did was in regard to another sort of kitten altogether.

It happened one year that there was a new baby in the house—a real two-legged baby this time. Zinnia of course knew about the baby; cats know all about everything. But she had never shown any great curiosity about it, possibly because she was busy herself, just then, bringing up a family of her own tucked away in a basket on the porch.

It was pleasant spring weather and the baby, who was just a few weeks old, was put out every morning to take her nap in a wicker cradle on the lawn. Usually she slept soundly, but one morning she happened to be a little restless and fretful. Now the crying of a young baby does sometimes sound rather like the mewing of a kitten, especially if the baby is sleepy, and just complaining to itself, as this one was.

Zinnia was seated on the doorstep, washing her face, and when she heard this peculiar sort of noise she pricked her ears up and strolled over to see what it was all about. Standing up with her front paws on the edge of the cradle she peered in, and it evidently didn't take her long to make up her mind just why the baby was crying. "What that child needs," thought Zinnia, "is a mouse!" So off she trotted down the garden, and round to the barn, to reappear a moment or so later with a nice fat mouse dangling from her mouth. Straight over to the cradle she went, and dropped the mouse in, with a little crooning mew, just to draw the baby's attention to it.

Now the baby's mother had watched all this, and she thought at first that Zinnia had just made a very funny mistake, and had dropped the mouse there by accident.

So she picked it up by its tail and took it over to Zinnia's own babies in their basket. But Zinnia would have none of this. She fished that dead mouse right out again and carried it over to the baby's cradle a second time, dropping it in with a little thump, as though to say: "You may think you know a lot about babies, but I've had more experience than you and, believe me or not, what that child needs is a *mouse!*"

Zinnia is a grandmother now, many times over. Most of her various babies are grown up and have families of their own, though each summer there are new ones to take their place.

Zinnia herself is growing middle-aged. Her little pointed face, with its pale-green slanting eyes, is a shade more pointed than it used to be; her black fur, always a little reddish in the light, is taking on more and more of a rusty tone, the color of an old iron kettle that has been lying out in the sun. But she is as keen a mouser as ever and her sense of responsibility is, if anything, more marked as time goes by. When I visited her last, only a short time ago, we had a very characteristic example of it.

A grown-up daughter of Zinnia's, Topsy, had died quite suddenly after only a couple of days' illness, leaving two pretty little kittens, about ten days older than Zinnia's own kittens. Luckily they were nearly big enough to lap for themselves, and meantime could be fed warm milk with a medicine dropper, but they missed their mother sadly, and the warmth and comfort of her body curled up beside them in the basket. Everyone was worried about poor Topsy's kittens and how they would get along without her, but before twenty-four hours had passed Zinnia, as usual, had come to the rescue. She heard the little orphans mewing, took one look at them, and without more ado seized first one and then the other and dragged them up two flights of stairs to the attic, where she settled them comfortably in the box with her own three babies.

There will be no neglected kittens in any house where Zinnia lives, as long as she is there to look after them.

VISITORS

I can remember, when I was quite a little girl, my mother asking me one evening to fetch her a clean handkerchief from her bureau drawer. I went upstairs—it was late summer dusk—and pulled open the drawer of the tall mahogany chest, and as I did so, peering inside, I saw two shining eyes staring at me from far back in the drawer among the piles of clean linen; very tiny eyes, pale green and brilliant like flames. At first I thought it was a mouse, and I clapped my hands to shoo it away, but the eyes never stirred. At that, feeling rather frightened—for I had no idea what this strange creature could be—I ran downstairs to call my mother, and when she brought a light, and the drawer was opened, we saw that the eyes belonged to a beautiful little pale gray moth, which must have found the drawer ajar in the daytime and crept in there to hide from the light.

I don't know if all moths' eyes shine in the dark, but I have often seen the same thing since, and the effect is very strange. It is their tinyness and their stillness that give one such an odd sensation, like gazing into the eyes of an elf.

If you have ever driven along a country road at night you may have seen the eyes of a rabbit, or perhaps a prowling fox, shining back at you from the road edge as the headlights of the car strike them. Going out with a flashlight after dark I have seen the eyes of a baby fieldmouse, two little pin points of fire, staring back from the grass on the lawn. The eyes of my little moth shone just like that, only greener and more jewel-like.

There is something even more beautiful about a moth than about a butterfly; the mysterious feathery softness of its wings, its furry head and delicately plumed antennæ. Perhaps the loveliest of all in our northern countryside is the large pale green lunar moth, as big as the palm of your hand and with all the magic of moonlight in its swaying silvery wings. This moth is apt to be a rare visitor, though in some places fairly common. If you look closely you may sometimes find one in the daytime, asleep and clinging to the underside of some leafy shrub, where its color makes it almost invisible. But I always feel sorry when I come across a lunar moth by daylight; it seems more than any other moth so essentially a creature of the night, belonging to the moon after which it is named.

I remember one evening in the country when the talk

happened to turn upon lunar moths, and we were saying
what a long time it was since we had seen one, and how
beautiful they were, when someone in the room said:
"Why, look!" And there outside the window were a pair
of lunar moths clinging to the wire screen, their great
pale wings fanning the air gently for an instant before
they rose and flew away.

As unexpected a visitor—and a much queerer one—
was the large praying mantis discovered performing her
devotions before the mirror in a third-floor New York
room. She had evidently arrived by way of the open window, but how so large an insect (she was over three inches
long) managed to cross the room unnoticed is a mystery.
For a creature of such sluggish aspect, the mantis certainly manages to get about with surprising ease. There
she was, surveying her own reflection, and what she could
see of the room, with an air of great interest. This insect
gets its name from the habit, when startled, of rearing up
from the waist and clasping its two forelegs together, but
the gesture suggests a threat rather than a prayer, as indeed it is. Those innocent-looking folded forelegs are
armed with sawlike blades and sharp hooks, and woe to
the prey that gets in their savage clutch! The mantis is
one of the very few insects also that can turn its head
independently of the body, and its way of doing so, staring

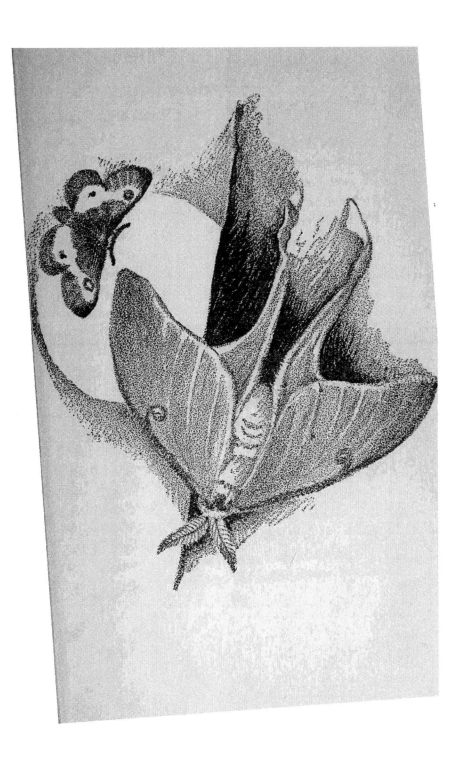

at you over its shoulder with those strange goggle eyes, is very disconcerting.

This one was coaxed on to a lead pencil and so carried to a geranium plant in the window box, which seemed a more fitting perch. But in a very short time there she was back again, this time crawling up the edge of a door, evidently not having seen enough of the room on her first visit. The pencil came into play once more, and this time the lady was dropped on the branches of the ailanthus tree in the yard below, with strict orders not to return.

Another creature which loves to come indoors, and is always turning up in most unexpected places, is the katydid. I think myself that katydids must be given a course in the study of human beings, and that there is perhaps a Katydid College, which sends out special pupils on a traveling scholarship. One seldom sees a katydid out of doors, probably because their color so nearly matches the fresh pale green of the leaves as to make them invisible, but often of an evening in the country you will hear a little friendly chirp, and there is the katydid, sitting on the table in the lamplight, looking very pleased with himself, or crawling in his peculiarly helpless way across the back of your chair.

The katydid is so pretty and so foolish-looking that one cannot help but like him. He seems always bewildered, always apologetic. Nature has given him an apparently serviceable pair of wings, and hind legs almost as long and powerful as those of his cousin the grasshopper, but they are quite wasted on him; he hates to use either, and much prefers crawling about in his own awkward and unsteady fashion. The katydid chorus, evening after evening, can become quite a pest, but how sorry we always feel when the last faint chirp is finally silenced, and we know that frost is close upon us!

A friendly little voice, too, is that of the cricket. There are few country hearthstones where a cricket will not sooner or later take up his abode. He keeps out of sight, hidden in some warm crack, but you hear his little persistent chirp every evening. One such cricket is a pleasant companion in the house, but too many of them are far from a blessing, for they will eat large holes in any article of silk or wool that they find lying about, and have a special fondness for silk stockings. Grasshoppers have the same bad trick, and if you ever spread a silk scarf to dry on the lawn, you will be very likely to find it riddled with holes like a sieve.

The Japanese love crickets, and make tiny straw cages for them to sit and chirp in. Perhaps they have found that the safest place to keep a pet cricket is in a cage, where he can do no damage.

66

One rarely hears of pet insects in this country, but a young friend of mine named Weda once had a pet butterfly and told me a very pretty story about him. Weda loves all animals, and when she found this butterfly, a large yellow swallowtail, one morning after a thunder shower, lying drenched and helpless in the wet grass, she brought him indoors to see if anything could be done for him. He had evidently just crawled out from the chrysalis and been caught by the shower before he had time to dry and spread his wings. They were crumpled, and one seemed injured, for he dragged it as he crawled.

Weda made him a cotton-wool bed in a candy box, with some green leaves round it, and left him to rest. When he seemed a little recovered, and happier looking, she offered him some sugar and water in a teaspoon. The butterfly stretched his long curled tongue out and began to suck the syrup up eagerly. The food gave him strength, and in an hour or so he was able to flutter about the room, awkwardly, for one of his wings still remained crumpled, though he managed to use it. He seemed quite tame, and would settle to rest every little while on her shoulder or arm, as though he understood that she was his friend and had helped him.

For a day or so she kept him indoors, where he seemed quite contented, and whenever he was hungry he would come and drink his sugar and water from her hand. He could distinguish Weda from other members of the family, and would always flutter about her when she came into the room.

When he grew stronger and could fly more steadily, Weda thought it was time to give him his liberty, so she carried him out into the garden, fully expecting him to spread wing and soar away. And so he did, but the strange part is that after several minutes of freedom, circling here and there in the sunshine, back he came again to flutter round her head and settle on her shoulder.

And so for several days he lived, flying happily about the garden and the tree tops, but always ready to greet her every time she came out. Even at a distance Weda could recognize him by his crumpled wing and uneven flight, and he always knew her. But a butterfly's life is a short one at the best; there came a morning when she went to look for him, and he was not there, and after a little search she found him lying dead by the edge of the flower border. Up to the very last he had not deserted the garden, nor the friend who had taken such care of him.

This is the only instance I have known of a butterfly being tamed, but perhaps other people have had a similar experience.

68

SPOT IN THE CONGO

The memory of Spot goes back to very early days. He came to us as a tiny puppy, white all over except for one black eye and a patch of black on his back. A baby fox-terrier puppy is about the nicest of all puppies, until he begins to grow up, and then usually nothing can beat him for mischief. Spot was no exception.

He chewed everything, slippers, hats, the legs of furniture. He would lie down on one's skirt—it was the day of long full skirts that trailed to the floor—and curl up in a tiny black and white ball, apparently sound asleep and dreaming, but when one got up, there would be a hole in the fabric almost big enough for him to crawl through. He would steal under the table at meal times, drag a napkin off someone's lap, preferably choosing a guest who was not on guard against his tricks, and eat the middle of it right out. And he had a perfect genius for stealing food out of the oven. If a dish of cutlets was set there to keep warm for the table, Spot would slip his paw cautiously in through the crack and hook them out one at a time, and when the dish was served there would

69

be two cutlets instead of five or six. And oddly enough
he never once burned himself. On one occasion he succeeded in stealing a leg of lamb almost as big as himself,
and dragged it off proudly to the end of the garden before he was discovered.

Like all dogs he loved to go for walks, but if by any
chance he was left alone in the house, instead, he racked
his brains to think of some mischief in revenge. He would
pull the table cover off, throw all the sofa pillows on the
floor and chew them, or upset everything else within
reach. Once he was shut in the kitchen, which seemed a
safe enough place. On our return the kitchen door refused to open. Upon pushing it there was a strange
rustling of paper, and we found that Spot had discovered
a big stack of newspapers, put away in a cupboard for
the ragman to take, and had torn every one of them into
shreds, till the kitchen floor was ankle deep in torn-up
paper.

One of his favorite tricks, which he learned himself,
was to sit on top of the garden gate. It was a fairly high
wooden gate, and solid, so that he could not see through
it, and the only thing to do, when he wanted to see what
was going on out in the street, was to climb up and look
over. So he would take a running jump, hitch his front
paws over the top and hang there, his hind toes braced

against the wood, and there he would perch contentedly by the hour, in what must have been a most uncomfortable position, but quite happy as long as he could stare up and down the street and bark at the other dogs going by.

Another performance of his—and about the silliest thing any puppy ever did—was to put his own tail out of joint. He sat down backwards one day, rather suddenly, only to arise with piercing yelps and begin galloping round the room in circles. For a moment no one could make out what the trouble was. We thought he had been stung by a wasp, until we noticed his absurd little tail with a crook where no crook should be. It took only a second to straighten it out with a quick pull, and then the expression of surprise and relief on Spot's face was very funny. This first accident must have weakened the tail, for after that he used to put it out of joint time and again in the same way, when he was romping. I don't think it ever hurt quite so much as the first time, but he would come running up, holding his tail at that same unnatural angle, yelping to say: "Straighten it out, please, *quick!*" And as a result of his tail being dislocated so often, there was a permanent kink in it, halfway down, which lasted all his life.

When Spot was about six months old the family went to live in Paris, and of course he came too. Paris delighted

73

him; he was quite as excited about it, and about the whole journey, as the children were. I think he really enjoyed the noise and bustle, and the unfamiliar sights. There was no garden gate to lean over, but there were a thousand fascinating things going on in the street which he could watch from the window, or when he went for walks. Above all things he loved riding in cabs; especially the old-fashioned open *fiacres* drawn by horses, of which there were still a number on the Paris streets. He liked to sit up with his paws on the side of the carriage, where he would look proudly about him while the coachman cracked his whip.

From being taken out to drive so often, and brought home again the same way, he must have reasoned quite clearly in his own mind that a cab was something that you could get into, at any point in the city where you happened to be, and which by some sort of natural process would always bring you safely home again to your own door.

We were out walking one day and Spot, who could usually look after himself in the street and so was seldom kept on a leash, managed to get lost. As soon as we missed him we called and whistled, but no Spot appeared. Once or twice before he had given us the slip like this, but always came back when he heard our whistle. This time, however, it seemed serious. We retraced our steps, looking for him everywhere.

But there was no sign of Spot, and the children began to be afraid he was lost for good this time, or that someone had stolen him, when finally, quite some distance from the point where we had first missed him, a workman who had watched us searching came up and said: "Are you looking for a little white dog with a black patch on his eye? Because I saw him just now, over there by the cab rank, getting into a cab."

The children hurried to the cab rank, and sure enough there was Spot, climbing up into one of the *fiacres* with a very determined air and barking indignantly, while the driver was just as determinedly trying to keep him out! Evidently he had decided that as he had lost us completely, the most sensible thing was to take a cab and drive home by himself.

He looked relieved to see us again, but was not quite so pleased when we made him climb down from the cab seat and walk home on his own four feet.

After about a year we gave Spot to some French friends who lived in the suburbs of Paris and had a large garden. They had taken a great fancy to him, and he to them, and we thought the dog would be better off there than in a city apartment, the arrangement of course being that if he seemed homesick or did not settle down happily, we should fetch him home again.

There are two kinds of dogs, those who can change owners easily and those who cannot, Spot belonged to the first class. He made friends readily, and while he was fond of everyone in the family he had no particular attachment to any one of us. After he left we had very good reports of him, and as he seemed perfectly happy we did not visit him for several weeks, in case the sight of his old family should unsettle his mind.

When we did go, the meeting was a very funny one.

Spot was completely transformed. He had become a French dog.

He looked smarter than I had ever seen him. His coat was shining; his plain English leather collar had disappeared, and instead he wore a gay affair with bells and dangles and a bow of red ribbon tied to it in our honor. Strangest of all, he had in those five weeks completely forgotten English!

He was delighted to see us, but when spoken to in English he could only stare with an utterly stupid expression. I never saw a dog look more embarrassed, and I seldom saw anything funnier than the look of relief that came over his face when his new owner addressed him in French. *That* he could understand; his ears pricked at once and his tail wagged. But English, no; he had ceased to understand that language entirely.

When the time came to leave, I was curious to see if Spot would want to follow us home.

We all walked together to the corner of the street, and Spot came pattering gaily along. When we had shaken hands I turned and walked a few paces down the road. Spot stood staring, with one paw raised. It was plain that he did not know whether I expected him to come with me or stay with his new owners, and I think he was really anxious not to hurt my feelings in the matter. But when I said cheerfully: "Well, goodbye, Spot!" his face brightened at once; he wagged his tail, came and jumped up to say goodbye, then turned and pranced off home without a single backward glance.

That was almost the last I saw of Spot, for not long after his owners moved to the south of France, where Spot had a chance to air his new French manners daily on the promenade at Nice, which I am sure he thoroughly enjoyed doing. But we had news from time to time, and one story that we heard seemed particularly characteristic of him.

His new master, a civil engineer, received an appointment to go to the Congo, in Africa, to superintend some work that was being done there, and he took Spot with him. Spot had always adored traveling; he made friends with everyone on the ship, and on arrival took to his strange tropical surroundings with as much enthusiasm as he had shown for every other new place he had ever visited.

Among other duties Spot's master had to make a trip up river, once a month, to report upon some constructional work that was in progress there. It was a journey of some three hundreds miles in a small steamboat, and it took altogether about twelve days. Spot loved these trips, and when the little boat touched at the landing once a month he was usually the first on board.

It happened one month that his master did not make this trip as usual. He had other matters to attend to. But this made no difference to Spot. When the regular morning came round, and he heard the little steamboat whistling, down he ran to the landing and hurried on board all by himself.

It was not until later in the day, some hours after the steamer had left that Spot's master missed him. At first he was worried, for the jungle is no safe place for little dogs to wander off alone. Then he remembered that it was steamboat day, and hurried down to the landing. Yes, the little white dog had been there; someone had seen him take the boat and go off.

78

When the steamer docked on its return journey there was Spot safe and sound, barking a shrill welcome from the deck. He had made the trip up river, spent his three days ashore, doubtless inspecting everything as usual, and taken the boat home again like the experienced traveler he was.

When I heard this story I seemed to see again the busy street in Paris years before, and a little white dog, calm and unflustered, climbing all by himself into a cab to drive home.

JIM CROW

Some relatives who lived in the country once had a tame crow, who, like most country crows, was called Jim. They had brought him up from a nestling; he had never been caged, nor had his wings been clipped, so that he had the freedom of the whole place, just like the dogs and the cats. He preferred, however, to spend most of his time about the dooryard and the house, where there was usually something going on to interest him.

Jim liked to help in everything, planting the garden, weeding, mowing the lawn, and often he was far more hindrance than help. He loved shelling peas or beans, and would always take the empty pods, carry them off one at a time and lay them in a pile. He learned to shell peas himself, holding the pod down with one foot and splitting it open with his bill.

There was a broken floor board on the back porch, and here Jim used to hide all his treasures, poking them down through the hole. If one of the dogs or cats went near this cubbyhole he would be driven off promptly, with indignant squawks, but if a member of the household strolled towards that end of the porch, then poor Jim would get dreadfully worried and uneasy, and do everything possible to divert his attention. Usually they would only pretend to look into this hole, just to tease him, but once in a while, when there seemed to be an unusual disappearance of small objects from about the house, or if some particular thing was being searched for, then the broken board would have to be pried up, and poor Jim's little treasure hoard laid bare, making him very unhappy and embarrassed.

Besides bean pods, bits of string and pencil stubs, there would be all sorts of things the existence of which perhaps had been almost forgotten, but which at some time or other had caught Jim's envious eye; a thimble, a little glass knob belonging to some bit of furniture, a fancy cigarette holder, oddments of nails and screws, ends of colored wool, and usually some small change as well. Then that particular object for which the hunt had been organized would be pounced on triumphantly, the board replaced, and the hoard left undisturbed till next time, while Jim, perched on the porch rail, flapped his wings and cawed dismally.

Jim was never spiteful but he was a great tease, and would make the cats' and dogs' lives a misery to them by tweaking their tails, shouting in their ears and pretending to steal their food.

There was an airedale in the family named Mike, who
had been blind for many years, ever since he had dis-
temper. The farm was high on a hillside, there was no
passing traffic, so Mike was perfectly safe, and although
blind he had learned his way about so well that he could
wander anywhere about the house and garden and knew
the position of each rock and tree. He had been taught
that dogs must not walk on garden beds, so whenever
he felt soft, fresh-raked earth under his feet he would
paw it a moment, to make quite sure, and then turn carefully and back away. He could pick his way quite easily,
turning out to avoid an obstacle just before he came to
it, and as everyone was careful, on his account, not to
leave any large object like a wheelbarrow or a bench
in any unaccustomed place, he very seldom bumped into
anything. In fact, a stranger seeing Mike for the first
time walking about the grounds would never have known
that he was blind at all, unless perhaps for the way he
would stand and listen, very carefully, with his head on
one side, if one spoke or called to him.

Mike could fetch and carry, and he loved having stones
thrown for him; a ball or stick was no use, because it
made no noise in falling. He would prick his ears, listening for the exact sound of the stone as it fell, and with
only that slight thud to guide him would rush straight

off and pick it up. If he did not get it the first time, then he would nose about in the grass and find it by his sense of smell. Sometimes he brought the wrong stone by mistake, but very seldom.

Being blind, and also very good natured and trustful, poor Mike fell an easy victim to Jim's tricks. Jim liked particularly to hop up very quietly and give a sudden tweak to Mike's toes as he lay asleep, and then caw with delight when Mike started up and barked. Sometimes he would keep this game up for so long that Mike really would get cross, in which case he laid his head down on his paws and sulked, a sign to Jim that the game had gone far enough. They were actually very good friends, for Mike was always gentle and although Jim teased him so much he was really fonder of him than of any other animal on the farm. And on chilly mornings he would spend hours perched on Mike's back as he lay sleeping, warming his toes in the dog's long shaggy coat.

Jim never flew far alone, but he liked going for walks. When any of the family set out for a walk he would first watch till he saw in which direction they were going, and let them get some distance. Then he would sail along to overtake them, swooping down as he flew quite close to their heads with a loud caw, and alight on some tree or fence post a little farther on, where he would

85

wait till they had passed and then catch up with them again, and this he would repeat for miles. And as the other pets liked going for walks too, the procession usually consisted of a dog first, leading the way, then someone of the family, and generally a cat or two tagging along in the rear, and as Jim flew past each one in turn he would swoop down and caw at them.

When Jim was about two years old the family moved to a town some distance away. Uncle Ted stayed behind for a couple of weeks alone, to see to the closing of the house, and various other things. One of his tasks was to build comfortable crates in which the cats—and Jim—could be shipped down to their new home. The dogs had already gone on ahead.

One trouble about a pet crow is that he cannot very well be left to shift for himself, however accustomed he is to complete freedom. The wild crows will seldom accept him again as one of themselves and though he can pick up plenty of food he is likely to be lonely, and there is always the danger that some farmer may shoot him by mistake for a wild bird. So it was thought best to bring Jim too, although he would probably have to spend a good part of his time shut up.

It was clear fall weather, just turning a little frosty at night, and Uncle Ted did most of his carpentering out

86

of doors, while Jim hopped about, watching him with the greatest interest. The cats had traveled before, and knew what all these preparations were about.

Jim had never seen a cage in his life, but he too seemed to know by instinct what it was. He watched the crates being built, with their barred sides, and he watched the cats trying them on, so to speak, being measured each in turn to make sure their particular crate would be roomy and comfortable. And in his own mind he must have put two and two together.

He had always been so tame that he would fly down at a call, and anyone could pick him up or pet him as he hopped about the garden. But not now. With the driving of each nail he seemed to grow more and more wary, and more and more suspicious, until when it came to the building of the last crate, his own, it was impossible to lay a hand on him. He hopped about, very inquisitive but always just out of reach, and by the time the last nail was driven home he took to the trees, and refused to come down to the ground at all. There he perched, staring at the crate with his head on one side and cawing derisively at any attempt to coax him down.

He had guessed only too well whom that last crate was for, and he was not going to be caught napping.

So in the end there was nothing for it but to leave

87

him there alone to keep house by himself, with a bag of corn in the barn where he could reach it easily through the open window.

When spring came again the corn was all gone, and so was Jim. Very likely he had been clever enough to use that wealth to make peace again with his old friends and relatives in one grand feast, and had then retired to end his days in honor and glory as a crow millionaire.

A BLACK-AND-WHITE BURGLAR

One summer we were living in a little camp near the Catskills. Beside the cabin there was a big wooden-floored tent which served as kitchen and living room, and in which the boys used to sleep. Their beds were on either side near the doorway, and they generally slept with the tent flaps looped back.

I particularly remember this wooden floor because it was underneath this one night that People, our cat, staged the biggest fight of his life with a huge gray farm cat who was his special enemy. They rolled and scratched and heaved about until the boards shook, all the furniture and crockery began to dance and we thought that an actual earthquake was taking place under our feet. It was not that night however, but another, that the boys were roused by a queer pit-pat on the floor. Outside the moon was shining but within the tent it was shadowy, so they could not see at first who their visitor was. The pit-pat continued busily, here and there, and presently something reached up and gave a little tug to the bedclothes.

Both the boys now were wide-awake, and peering out

from under the blankets they saw, in a ray of moonlight that came through the tent flap, a large black and white skunk. He went to each bed in turn, standing up with his forepaw against the side, and staring very inquisitively at the occupant. The boys kept quite still, neither dared to move or shoo him off, for fear they might annoy him, for an annoyed skunk can be a most unpleasant visitor, especially in a tent. But the skunk evidently satisfied himself that they were both sound asleep, and off he trotted to the kitchen end of the tent, where they could hear him rummaging about among the pails and saucepans, trying to find a scrap of stale bread or something else to eat. Something he did find, for they heard him munching, and then after a little while he trotted back and disappeared again into the night.

After that we learned to expect our little burglar nearly every moonlight night, and as he never troubled anyone, or did any harm, we left scraps of food where he could find them as a reward for good behavior. Cold potatoes he was especially fond of. We found out that he had a hole under the floor of the cabin, where he had been living very comfortably for a couple of years. He had evidently made friends with the cat, who being also black and white, and about his size, may have seemed like a sort of cousin, for there was never any trouble between

them. People would sit on the bench outside with his paws tucked under and watch him with interest as he pattered about. I think he even slipped him advice as to what had been left over from supper, and where.

Cats and skunks, so far as I can make out, generally get along together very well, both being quiet and peace-loving animals. A dog is a different matter, and the only times we were made unpleasantly aware of the skunk's presence, by that pungent peculiar odor which will wake one up from the deepest sleep, was when, on two occasions, a wandering farm dog came prowling round to disturb him.

Personally I don't a bit mind the smell of a skunk so long as it is not too close. If a skunk lives anywhere near by the most one will notice is a slight whiff now and then, usually on moist or rainy nights, which seems always as much a part of the country as the smell of woodsmoke from autumn bonfires. Close to, it has a strong suggestion of creosote, and is as irritating to the nose and eyes. It is the skunk's only weapon of defense—and quite an effective one—but he only uses it when seriously alarmed or threatened.

The first live skunk I ever saw was many years ago, on a visit to a farm. We had looked at all the calves and

93

the pigs and the horses, and then one of the boys said: "I've got something else to show you, out here."

I followed him to an outhouse and there in a wired pen were three baby skunks, just a few weeks old. They were black all over and looked not unlike young kittens except for their pointed noses, and they were tumbling and playing about in the straw just as kittens might. They had been born in the outhouse, where their father and mother also lived, and the whole family were perfectly tame. I saw the father, a very handsome full-grown skunk, in a pen by himself, poking his nose through the wire for a scrap of bread, but the mother skunk was curled up out of sight in her sleeping box and my friend said it would be as well not to disturb her, as she didn't always care for strangers, especially when she had babies about, and might resent our coming in to look at them.

The boy who owned these skunks had caught them when they were young and had had them for nearly two years, and he told me that he had kept a great many pet skunks and that they were very easily tamed and always well behaved. "Once they get to know you," he added.

I know of a house in the South where for several years a skunk has brought up a family every summer under the porch floor. And every morning, while the members of the household are having their own breakfast on the porch, in the shade of the honeysuckle vine, she marches proudly out leading her babies, four or five of them in single file, and brings them right up the steps to where a bowl of bread and milk is always set ready for them. House guests who have not been warned beforehand are very surprised to see the little procession come marching up, plumed tails waving, to take their places round the breakfast bowl.

Mother skunks like to take their babies walking, usually in single file and at a slow and dignified pace. Sometimes one may see the whole family crossing some quiet country road, and taking their own time about it. And they are such good-looking youngsters that no wonder Mama is proud of them.

A dog who has had one serious encounter with a skunk will generally leave them alone thereafter, and do his barking and yapping at a safe distance. Little black Trotty who was so inquisitive about the woodchucks was also punished once for her curiosity about a certain strange animal—strange to her—who lived in a patch of underbrush by the corner of an old stone wall down near the brook. Whenever we went for a walk in that direction, towards evening, Trotty would break away to explore that corner, and we would hear her voice raised in shrill yelps of excitement, without however paying much attention to it, for Trotty is given to excitement over trifles, especially in the woods. But one evening the yelping was shriller and more jubilant than usual. "Trotty has started a rabbit," we said, and strolled on, leaving her to the chase.

A little later, as we were all seated on the porch, a strong and penetrating odor filled the air. Someone exclaimed: "Skunk!" and we looked here and there to see where the skunk could be. All we could see was Trotty, rolling on the grass and doing her very best, it seemed, to bury herself under the ground, and all the while sneezing and rubbing her little black nose piteously with her paws.

Trotty had met her first skunk face to face, and she didn't at all like it. Neither, for that matter, did anyone else. For once in her life Trotty was a thoroughly unpopular little dog. It was no use looking miserable and trying to climb into everyone's lap in turn; no one wanted her. She stood, crestfallen and unhappy, while we rubbed her off with sacks, and then led her away and shut her in to spend the night alone in the barn.

Somewhere down in the hollow the skunk was doubtless thinking: "Well, I've taught *that* nosey little person a good lesson, anyhow!"

SPIKY AND CO.

For a long while we had forgotten all about hedgehogs, and then one day someone happened to mention them. Could one still buy a hedgehog nowadays, or had they gone out of fashion altogether? It seemed a question that needed to be settled then and there, and so the very next morning I set out on my quest. Luckily we were living in London at the time, and if a hedgehog—or almost any other queer pet, for that matter—can be found at all, it can be found in London, if you know where to look.

Perhaps very few of you who may happen to read this book have ever seen a hedgehog, unless it was in Europe, where they are common enough. A hedgehog is rather like a small porcupine, except that he has no tail, and that instead of long handsome quills he has stiff prickly bristles set closely all over his back, much like magnified prickles on a chestnut burr. They are quite sharp too, and though usually he carries them smoothed down, he can raise them on end in an instant if alarmed. And if that is not sufficient defense he will then curl himself up into a tight ball, head and feet tucked out of sight and nothing but spines showing, and you may pick him up or roll him about as much as you please, nothing will induce him to uncurl again till he thinks that all danger is past. Then gradually a small pointed nose will stick out, two bright eyes peer this way and that, four little feet appear as if by magic, and your hedgehog trots away.

But why describe a hedgehog? You can see his portrait in the pages of *Alice in Wonderland*.

The hedgehog knows very well that once he is curled up tightly no dog can harm him and that even human beings are likely to let him alone, for his spines are so sharp that he is most unpleasant to handle, and quite impossible to unroll.

In England many years ago a number of the old-fashioned houses had stone-paved kitchens and sculleries. Cockroaches, or "black beetles," as they are called there, infested many of these old floors, living under the stones and in the warm cracks near the brick fireplace, and if you were troubled with black beetles in your kitchen the remedy was to buy a hedgehog and turn him loose there. Hedgehogs live largely upon insects, and are nocturnal creatures; all day long your hedgehog would sleep, hidden away in some corner, by night he would wake up and prowl busily about for his supper, and in a week or so every single beetle would have disappeared. Then, if it was warm weather, you usually turned him out into the garden to eat up the worms and snails and slugs there.

It did seem, as I toiled from one London bird shop to another, that hedgehogs had indeed gone out of fashion. No one sold them. But at last I found a tiny shop tucked away under the railway arches in Camden Town, kept by an old man who at least seemed to know about hedge-hogs and to take some interest in them. He promised to get me a hedgehog by next week, and he was as good as his word. He found not only one hedgehog, but two, and wanted me to take them both. "You'd better have two," he said. "You never know!" But one hedgehog seemed quite enough for the time being.

He dropped the queer spiny ball into a brown paper bag and I carried my hedgehog home. It was early spring; he must have been just routed out from his long winter sleep in some bank or hedgerow and was quite cross about it, for he refused obstinately to uncurl. He was dingy and earthy, with dried leaves and mud caked to his bristles, and the first thing was to give him a bath, before one could even see what he really looked like. Tepid water and a scrub-brush worked wonders; the dirt fairly flowed from him, and at last he uncurled in sheer fury, opened a beady eye and began to scramble indignantly up the side of the tub.

For many days he remained shy and surly. When he thought he was not observed he would uncurl and poke his pointed nose out long enough to take stock of his new surroundings, but a step or movement would send him back again into an obstinate bristly ball. Then he gained confidence, began to explore the house a little, take an interest in the legs of the chairs and the strange unaccustomed feeling of the carpet under his toes. And before a week was over he was trotting busily about, sniffing here and there and staring around with his solemn little gray face, perfectly at home.

He lived on bread-and-milk and table scraps, put out for him in a saucer every evening, for hedgehogs prefer to feed after dusk. And I will say that he had terrible table manners, and made more noise eating than would seem possible for an animal his size. At intervals during the evening we would hear "gobble-gobble—glug-glug" from the darkened kitchen, like a whole troughful of little pigs. Only the hedgehog eating his supper!

In fact he was one of the greediest creatures I ever knew, perhaps because he preferred the kind of meals we gave him to the worms and slugs he had been used to. Even when rolled up asleep one could make him uncurl at any time just by holding a bit of cake near the spot one imagined his nose to be.

Most of the day he slept. Towards dusk he would wake up and begin to feel lively, and the night he spent trotting about the house. By morning he was rolled up fast asleep again somewhere, and there would be a hunt to find him, for he usually chose a new hiding place every time. Behind the sofa cushions, in a corner of the broom closet, sometimes in the pocket of an overcoat hanging on the hall rack, for hedgehogs climb anywhere. And more than once he was discovered among the warm ashes under the kitchen stove box. He was not always easy to see, but sooner or later one would put a hand on something sharp and prickly, stowed away where one least expected it, and there was the hedgehog.

"Spiky" seemed by far the most appropriate name for him, though as he grew tamer he learned to keep his bristles smoothed down most of the time, except when he was asleep, and so could be handled and even stroked without discomfort. He would come when one called him, liked to sit in one's lap, especially at tea time, when there were bits of cake and warm tea to be lapped from a teaspoon, and he was fond of having his nose scratched for him. But there was a funny trick about this.

He would stretch his head up with every sign of pleasure, and then, when he had had enough of the scratching, give a quick unexpected snap at your finger. Now a hedgehog's bite cannot possibly hurt anyone; his teeth are so tiny that the worst you get is a sharp prick, scarcely deep enough to break the skin, but his quick way of biting seems so ferocious that it startles one very much. So it became a favorite trick, when any new visitor became interested in Spiky, for some member of the family to say innocently: "Just scratch his nose. He loves to have his nose scratched!" And then watch for the surprise!

One day Spiky got lost; really lost, not merely mislaid as usual. He was nowhere in the house, he hadn't been thrown out with the rubbish, an accident which had happened once, but luckily he had been retrieved in time, before the dust man came along. For hedgehogs will go to sleep in the most unaccountable places and are quite apt to get mixed up in the waste-paper basket or the trash bin. Nor had he crawled into an overcoat pocket and been carried away unnoticed by the owner, which had also happened once, and the mistake been discovered only when the wearer was waiting for his omnibus at the street corner and put a hand in the pocket to find a handkerchief—finding the hedgehog instead, not at all a good substitute. This time he was really lost.

By now life without a hedgehog just seemed empty. I hurried off again to the little bird shop in Camden Town, and by great good luck found the second hedgehog still unsold. Now we knew why the old man had said: "Better take two!"

"Hedgehogs is outwitting things," he remarked now—evidently not surprised to see me back again so soon—as he dumped Spiky the second into a paper bag and twisted the top over, for all the world as if it held a melon or a bunch of carrots. "You never know where to lay your 'and on 'em! Sooner or later, they'll be off."

Spiky the second was duly carried home, bathed and in his turn introduced to the family circle. And—as usually happens when one is in too great a hurry to replace a mislaid article—no sooner was he fairly installed in the house than our original Spiky turned up, strolling casually across the tennis courts in the dusk. So now we had two, just exactly alike, and the only way we could ever tell them apart, after a few days, was that Spiky the first still snapped when his nose was scratched, and Spiky the second didn't.

With summer coming on, and doors and windows wide open, it became harder and harder to keep track of our pets. For no pen or wire netting will keep a hedgehog in if he really wants to get out. They will dig or climb, and squeeze their way through almost any crevice, and like tortoises they are born travelers. Spiky the first finally took to the tennis courts for good, where he was to be met at intervals, wandering of an evening round the shrubbery to look for snails, and Spiky the second went to live with a friend who had a walled garden, and wanted a hedgehog to eat up the worms and slugs in the strawberry bed.

About a year later, in Italy, we had another hedgehog. He was brought to us by the old man who mended the roads, and who found him curled up among the underbrush in a ditch. He was bigger than our English hedgehogs and much wilder; nothing would induce him to settle down and be friendly. We at last put him for safekeeping into an empty room among the attics, and shut him in. First he dug away at the skirting board with his strong little claws, and when that breach was repaired managed somehow to hoist himself to the window, and from there out to the wide rain gutter which ran around the roof. Then it was a simple matter—at least to a determined hedgehog—to squeeze through the overflow hole, roll himself into a ball and simply drop to the ground, where he could uncurl and walk away at his leisure, none the worse, for a rolled-up hedgehog is quite invulnerable to bumps—or to anything else for that matter—and those strong elastic spines serve to break his fall very conveniently.

As the old bird shopkeeper said: "Hedgehogs is out-witting things!"

COUNTRY NEIGHBORS

Outside my bedroom window, where on waking in the morning I have a clear view of its upper branches, is an old cherry tree. It is not the only cherry tree around, but its cherries are the sweetest and ripen the earliest. So it is an object of particular interest to all our small woodland neighbors.

The first visitor of all, about a week before the cherries begin to ripen, is a yellow-stockinged blackbird. He must come from some distance, for I never see him at other times, and he is not among the familiar birds about the garden. He arrives at dawn, alights with an important air on some upper branch, cocks his head and inspects the condition of the fruit very carefully, after which he will give a few jubilant squawks of satisfaction and fly off again, probably to report to his friends and neighbors how things are getting along. About ten days later—for he times his visits most accurately—he will be back again to snatch his first share of the fruit.

After the blackbird come the gray squirrels. They live in the thick woods behind the cottage, where they have been keeping their own close watch on the cherries. Just after daybreak one hears a great swishing and rustling in the tree tops, and they arrive in a troupe, all together, headed by what must be the great-grandfather of all the squirrels in the neighborhood, judging by his size, shouting, chattering, swinging from branch to branch till they reach the cherry tree. There they separate, and settle down each to his own breakfast.

There is a low extension roof, the ell of the cottage, which runs out just below the cherry tree, and after the squirrels have finished eating, which takes about ten minutes, for they settle down to the job with great speed and seriousness, they drop to the roof and go through their morning exercises. These consist of leaping, playing tag, and scampering back and forth over the shingles, five or six at a time, and making such a din that further sleep is impossible. Occasionally their leader takes a flying leap to the upper roof, and one can hear his teeth tearing angrily at the edge of the cornice.

For the squirrels have an old grudge against us, which

107

they have never forgotten. Before we came to live here the cottage stood empty for some years. It had fallen into disrepair; there was a broken knothole high up in one wall and through this the squirrels had made their way indoors. Winter after winter they had the whole house, and particularly the attic, all to themselves; it was splendid. They owned the best dwelling, from a squirrel point of view, for miles around.

And then we came along, and spoiled it all.

The knothole was repaired, the old roof torn off and replaced by new shingles, and when the attic was thus laid bare the workmen found, under the slope of the eaves, an enormous nest, the winter home of generations of squirrels. When it was all pulled out, armful after armful of shredded paper, hay, ends of rag and empty nutshells, there was enough to fill four large sacks! Every scrap of wool or paper about the house must have gone to the making of that nest, all except an old Bible and a school arithmetic, which for some reason they had not touched, though other books on the attic floor had been long torn to shreds.

There were no young squirrels in the nest; it was late in the season and the tribe had already taken to the woods, but still the squirrels were furious and I can't blame them. The old grandfather sat in the cherry tree day after day and scolded the carpenters at the top of his voice. He must have called them every name in squirrel language.

And still they are angry. Even though we offer them the barn, with a top window left open for them all winter, they feel it a poor exchange for their ancestral home. And when they sit in the cherry tree of an evening and throw their discarded cherry stones down on our heads, I suspect them of taking careful and particular aim.

Between the squirrels and the birds, orioles, blue jays and catbirds who throng the cherry tree when the fruit is ripe, there is a continual squabble. The birds chirp impudently; the squirrels scream and chatter back at them. While to and fro, under cover of the garden fence, scuttles a timid little shadow, the striped chipmunk, watching to snatch some half-eaten cherry that his bolder neighbors have dropped.

We human beings, of course, have no right to the cherries at all. Birds and squirrels alike hate us when the time comes for gathering the fruit. Still the best cherries always seem to grow on the very tips of the branches, just out of reach, so there are always plenty left.

Later on the red cherries ripen, farther up the garden, and then the squirrels move in a body to that tree, and we see fewer of them about the house.

One day while we were watching the troupe at play, a friend told me an amusing story about some other gray squirrels who had a nest in the garage behind her house. There were a mother squirrel and five little ones, and every day, when the babies were big enough to leave their nest, the mother would lead them out to sit in a row on the garage roof and take a look at the world. She played with them, taught them to run and jump, and to cling with their claws to the sloping roof-top. At last one day she decided that they were quite old enough to take to the woods and look after themselves. There was a tree growing near the garage, a few feet away, but quite an easy squirrel jump. So she began by showing them how simple it was to leap from the roof to the tree, run down the tree trunk to the ground, and then back again. She did this over and over, a number of times, calling her babies to follow, but the little squirrels were either timid or plain lazy; they preferred to sit in a row on the ridgepole and watch her. The mother coaxed them and scolded them; she would lead each one in turn to the edge of the roof and show him how it was done, but when it came to jumping the little squirrels always hung back. She tried to push them; they clung fast with their claws.

Then finally the mother lost her patience; she had

quite enough of such nonsense. So without more ado she seized one of the babies in her mouth, just as a mother cat carries a kitten, leaped with him to the tree branch and ran down to the ground, where she dumped him sternly in the long grass and went back for the next. And so in turn, till she had all five safe on the ground. And there she left them to fend for themselves, washing her hands of all further responsibility. She had done her duty by them, and that was that.

Squirrels and birds are not our only neighbors. There are a family of woodchucks down by the orchard who steal out in early morning to pick up the fallen pears and apples; cousins of the woodchucks who live on the pasture slope across the road. And in this same pasture are five cows, two elderly cows and three younger ones, who pay us a visit nearly every day. They climb the hill to spend the noon hour under the oak tree just opposite the house where, however hot the day, there is always a breeze from across the valley.

One of the older cows, the white one, has only one eye. She lost the other one fighting, some years ago, the farmer told us, and it must have been a good many years, for she is so gentle now that one cannot imagine her fighting anyone. That blind eye gives her a queer look; when you speak to her she always has to turn her head right around, before she can see who it is, and she is nervous of anyone approaching her on the blind side. She likes bread, and knows that every morning, when the bread box is turned out, there are sure to be a few stale crusts put aside for her. If I am not there on time she pokes her mild old face through the fence, and moos till I arrive.

If cows can talk they must find a great deal to gossip about among themselves. They take so much interest in everything, and I know that these particular cows watch all that goes on. If one comes to the doorway, or crosses the garden, five heads are turned instantly to stare; five mouths pause in their chewing, waiting till the interruption is over. They are for all the world like five gossipy old ladies grouped with their knitting under the oak tree, discussing their neighbors' doings.

Before the tree shadow begins to lengthen they will rise one by one, give a last look around them, and then start on their leisurely stroll down the big pasture to the foot of the hill, where four o'clock always finds them gathered at the bars, waiting to be fetched home for milking. In the country, cows are a pretty good guide to the time of day.

About now, if there is still a cherry or so left, the squirrels will be back for their evening meal, noisy as usual, running from branch to branch and calling insults to the catbird, who talks back to them. The chipmunk appears noiselessly from some hole in the stone wall, his striped coat shining in the afternoon sunlight. He is waiting for a chance to steal across the road unseen, being well aware that People, the big black cat, is curled somewhere under the lilac bush at the top of the rock garden, a favorite nook from which he can keep watch on everything that happens. People has never caught a chipmunk yet; he is scolded even for stalking them. But he once managed to play a fine trick on me.

It was at the seashore; there was a wooden path behind our house, leading out to the windmill, and here a number of chipmunks used to play. One evening, just as we were all sitting down to supper, someone cried: "People has caught a chipmunk!" I looked out through the window and saw with horror People, after all our threats and warnings, playing with a striped object on the pathway, shaking it, biting it and tossing it into the air, always with one eye on the window to see if he were being observed. I rushed out, hoping to be still in time, and People pretended to be much concerned at my appearance, the bad cat! But when I snatched his plaything from him it was nothing but a dried wisp of fur and bone—a chipmunk once, but dead these many months! People was

113

delighted with his joke, rolling over and purring round my ankles, and though I threw the mummified chipmunk away into the bushes, where he had probably found it in the first place, he took pains to unearth it again next evening at supper time, and go through the same performance once more in full view of the dining-room windows. But this time the joke fell flat; we were not to be fooled a second time.

People has his own little door in the cottage, through which he can go in or out as he chooses, night or day. He is as good as a watchdog, and any unusual sound at night will bring him out to investigate. When Trotty stayed with us nothing would do but she must use People's private doorway too, much to his disgust. It is quite a tight fit for a little dog—only seven inches square, measured just to accommodate People's whiskers comfortably —but she would squeeze and scrabble and finally emerge triumphant with a loud plop, just like the sound of a cork being pulled from a tight bottle neck!

It is getting on for sunset. Very soon People will come out from under the lilac bush where he has been snoring all afternoon, yawn and stretch his claws and saunter slowly down the rock garden, pausing by his own special catnip bush on the way. Then in through his little door-